# YOUR KNOWLEDGE HAS VALUE

- We will publish your bachelor's and
  master's thesis, essays and papers

- Your own eBook and book -
  sold worldwide in all relevant shops

- Earn money with each sale

Upload your text at www.GRIN.com
and publish for free

**Bibliographic information published by the German National Library:**

The German National Library lists this publication in the National Bibliography;
detailed bibliographic data are available on the Internet at http://dnb.dnb.de .



**Imprint:**

Copyright © 2015 GRIN Verlag, Open Publishing GmbH
Print and binding: Books on Demand GmbH, Norderstedt Germany
ISBN: 9783668292000

**This book at GRIN:**

http://www.grin.com/en/e-book/339506/reduction-in-cost-of-quality-by-using-robust-
design

**Jalal Khan**

# Reduction in cost of quality by using robust design

GRIN Publishing

# REDUCTION OF COST OF QUALITY BY USING ROBUST DESIGN: A CASE STUDY IN AUTOMOBILE INDUSTRY

Course : Manufacturing Systems Engineering
PREPARED BY JALAL KHAN

# Contents

# Learning Objectives

After completion of this session you will:

- Understand the relation between Robust Design and Cost of Quality
- Be able to identify the quality costs in an organization
- Understand the impact of measuring the cost of quality

# 1.  Introduction

- **Genichi Taguchi** defines quality as, "The quality of a product is the (minimum) loss imparted by the product to the society from the time product is shipped" (Bryne and Taguchi, 1986). Quality has been defined by many as being within specifications, zero defect or customer satisfaction.

- **Robust Design** is defined as reducing variation in a product without eliminating the causes of the variation. In other words, making the product or process **insensitive to variation**. This in turn increases the overall quality of a product making the product robust.

- "We need to communicate to management the impact of quality in language they understand which is often in terms of dollars."

- The present thesis proposes a systematic design framework for cost of quality that embeds Taguchi's method and other robustness criteria to reduce the overall COQ of a product.

# Continued...

❑ Cost of Quality:

- COQ is usually understood as the sum of conformance plus non-conformance costs,

    - where cost of conformance is the price paid for prevention of poor quality (viz. inspection and quality appraisal) and
    - cost of non-conformance is the cost of poor quality caused by ` product and service failure (viz. rework and returns)

❑ Robust Design:

- Robust Design improves the quality of a product by minimizing the effects of variation without eliminating the causes (since they are too difficult or too expensive to control).
- It is an off-line quality control method that is instituted at both the product and process design stage.

# Continued...

❑ Classification of COQ:

| COQ Models | | |
|---|---|---|
| **P-A-F Model** | **Crosby's Model** | **Opportunity or intangible cost Model** |
| • Prevention<br>• Appraisal<br>• Failure | • Prevention<br>• Appraisal<br>• Failure<br>• Opportunity | • Conformance<br>• non-conformance<br>• opportunity<br>• Tangibles<br>• intangibles P-A-F2 |

# Continued...



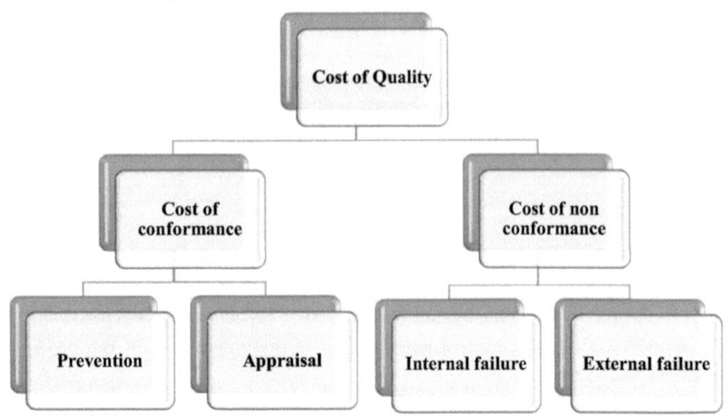


# Continued...

☐ Need for Robust Design:

- Robustness reduces variation in parts by reducing the effects of uncontrollable variation.
- Another advantage is that lower quality parts or parts with higher tolerances can be used and a quality product can still be made.
- This method is also good, because you are designing the robustness into the product and process instead of trying to fix variation problem after they occur.

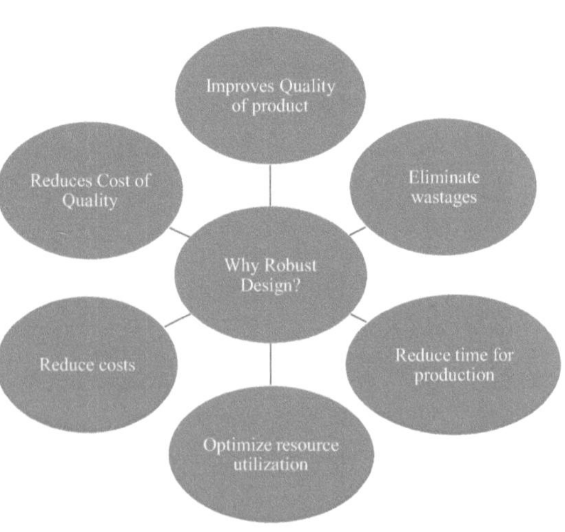

# Continued...

❑ **COQ and Robust Design:**

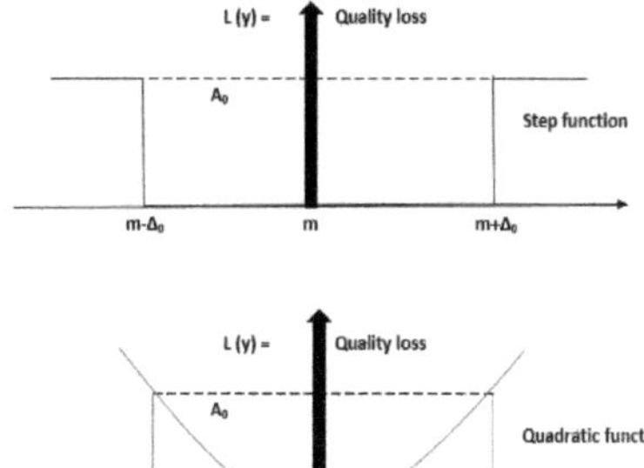

- Quality loss increases if the tolerance deviates from the target value 'm'
- Tolerance limit range is from m+Δ to m-Δ as shown in the figure.
- Robustness of a design reduces the tolerance limit range thereby increasing the quality of a product.
- Hence COQ of a product will reduce and optimum design can be achieved for the product.

# Continued...

❑ Overview of Industry:

- Bharat Gears Ltd. (BGL) is one of the world leaders in gears technology and India's largest gear manufacturer.
- BGL manufactures a wide range of Ring Gears and Pinions, Transmission Gears and Shafts, Differential Gears, Gear Boxes majorly for the automotive industry.
- Plants are located at Faridabad, Satara and Silphata.

- **Vision:** Key contributor to Self-reliant India and globally respected and sought after partner
- **Mission:** Helping India grow and a major supply hub for gear used in automobile sectors.
- **Values:** Precision, Productivity and Speed, Knowledge and Skill, Process driven, Innovation and Creativity, Collaboration and Team work.

# Continued...

❑ Gear Hobbing Process:

- Hobbing is a machining process used in BGL for gear cutting, cutting splines and cutting sprockets on a hobbing machine, which is a special type of milling machine.
- The teeth or splines are progressively cut into the work piece by a series of cuts made by a cutting tool called a hob.

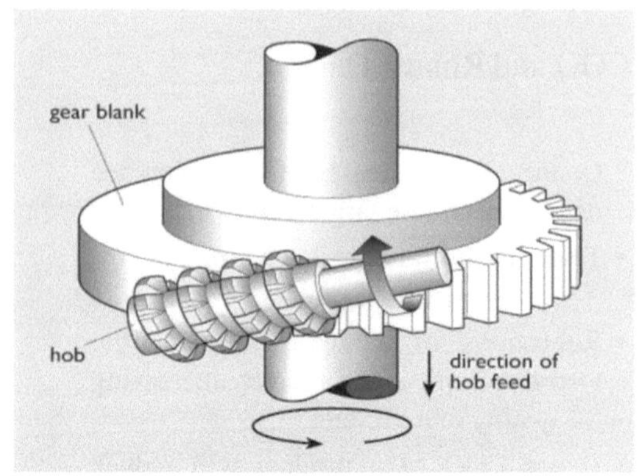


Source: http://www.ustudy.in/sites/default/files/images/gear%20hobbing.jpg

# Continued...

❑ Challenges faced by BGL:

Most of the products that BGL produces are gears that are used for transmission purpose in automobile. These gear products have some tolerance limit within which all gears are considered acceptable. This leads to:

- Acceptance of those products which were many a times of inferior quality but accepted due to the standard set.
- Increase in cost of quality of a product.
- The design were not robust and sensitive to various external factors.

To counter all these major factors, Robust Design approach is selected such that the product manufactured are nearest to the target value.

# Continued...

❑ Need for the research:

- The growing demand for advanced technology with minimum cost is the need of the hour.

- The current economic downturn, the rising cost of energy, raw materials, etc.

- Imparting role as a major player in global automotive market thereby maintaining its position as employer of many people.

- Need to develop products which are insensitive to external factors thereby making the product robust.

- Design of a system to decrease the cost of quality of a product and hence, saving money.

# Continued...

❑ Motivation and Objectives:

- To study the process and benefits of COQ and Robust Design as applied in the BGL with a special reference to transmission gear used in shaft.

- To optimize effectiveness of the processes.

- To study the role of COQ practices and Robust Design implementation for the selected industry sector and amendment of impact of the same on organizational performance.

- To develop a methodology for reduction in Cost of Quality in BGL by integration of Robust Design methodology.

- On the basis of analysis of result, identify areas of excellence and recommend some useful practices for the selected sector.

# Continued...

❑ Problem Definition:

- The process of gear manufacturing is carried out through hobbing machine. Due to various noise factors and its high impact, the quality of gear produced detoriated.

- Based on earlier production of gear, researcher had analyzed that many gears produced, though they met the specification, or in other words were within the tolerance limit, they still had an impact the overall quality of the product.

- Robust Design methodology is to be used and various designs were obtained by changing the various parameters. Finally, optimum design is selected and tested so as to achieve maximum quality of a product.

- In this research work, it is decided to work on the improvement of gear design by implementation of robust design methodology thereby reducing cost of quality.

# 2.   Literature Review

❑ Past Review on Robust Design and COQ:

| Sr no | Author | Title | Tools & technique | Remarks |
|---|---|---|---|---|
| 1 | Achamyeleh A. Kassie et al. (2002) [1] | Minimization of Casting Defects | This paper conducted taguchi's design of experiment at Akaki Basic Metals Industry (ABMI) to optimize process parameters in production of steel by casting process. | It was concluded that shrinkage defect can be minimized by proper design of gating system for different environmental and working conditions. |
| 2 | Adil and A. Moutawakil (2012) [2] | The Quality Cost Reduction in Hollow Glass manufacturing by Taguchi Method | This paper carried out an experiment to optimize the weight of a glass manufacturing bottle of a product called CC35 by Taguchi method of parameter design using L16 orthogonal array. | Experiment showed that the Taguchi method applied on the glass bottle's manufacturing process (CC35) ensures best stability and reduces the weight which results in the quality cost down to 92%. |
| 3 | Andrea Schiffauerova and Vince Thomson (2007) [3] | A Review of Research on Cost of Quality Models and Best Practices | This paper discussed and compared the quality programs of four major companies and explains the benefits of the eventual adoption of a COQ approach in each case. | It was found that the company that worked on COQ models were able to reduce the cost of quality significantly. |

# Continued...

| Sr no | Author | Title | Tools & technique | Remarks |
|---|---|---|---|---|
| 4 | B P Gautham, Nagesh Kulkarni and Janet K. Allen (2004) [4] | Minimization of Casting Defects | This paper carried out design of transmission gears for an automotive system with desired reliability, cost, and robustness using the AGMA (American Gear Manufacturing Association) standards for gear design. | They proposed method for robust design and eliminated the conventional factor of safety and reliability factor used in AGMA based design procedure. |
| 5 | Davison Zimwara and Lameck Mugwagwa (2013) [5] | Cost of Quality as a Driver for Continue Improvement - Case Study | The paper presented a significant step towards overcoming the difficulties In manufacturing firms by providing a systematic practical approach to addressing the cost of quality. | They assessed missed quality objectives and means for quantifying and implementing corrective actions for the same. |
| 6 | Dobrin Cosmin, and Stănciuc Ana-Maria (2006) [7] | A Review of Research on Cost of Quality Models and Best Practices | This paper discussed that an organization needs to adopt framework to classify costs and to focus on existing cost of quality (COQ) models. | They explained the various associated losses which even included losses such as consumer dissatisfaction, warranty costs to the producer, and loss due to a company's bad reputation, which leads to eventual loss of market share. |

# Continued...

| Sr no | Author | Title | Tools & technique | Remarks |
|---|---|---|---|---|
| 7 | Joseph Juran (1996) [10] | Quality Cost Analysis: Benefits and Risks | He explained the costs associated with preventing, finding, and correcting defective work. | It was concluded that key functions of a quality engineer is the reduction of the total cost of quality associated with a product. |
| 8 | Juliana Litecka, Tomas Horvat and Veronika Fecova (2007) [12] | Classification of Factors affecting Hob wear | The paper proposed a literature on increasing the life of cutting tool used for gear manufacturing. They dealt with factors which influence gear hob wearing and their effect on each types of hob wear. | It was concluded that the costs on the change of worn tool are more the more often the change is realized which ultimately increases the cost of quality. |
| 9 | K R Balachandran and Bin Srinidhi (1991) [14] | Target Analysis: Cost, Quality or Both | The paper explains the traditional design and robust design in detail. | The traditional view is that improvements in quality entail increased costs and are generally associated with reduced productivity which is vice versa in robust design. They eventually compared the differences between implementing the system in USA and implementing the system in Japan. |

# Continued...

| Sr no | Author | Title | Tools & technique | Remarks |
| --- | --- | --- | --- | --- |
| 10 | Mohd. Muzammil, Prem Pal Singh, and Faisal Talib (2003) [18] | Optimization of Gear Blank Casting Process by Using Taguchi's Robust Design Technique | The paper demonstrated optimization of the gear blank casting process by using Taguchi's Robust Design technique. The metal casting process involves a large number of parameter affecting the various casting quality features of the product. | A taguchi's experiment was performed using the levels obtained. The weight of the castings produced was found to have a value close to the target value. |
| 11 | N.M.Vaxevanidis, G. Petropoulos, J. Avakumovic, A. Mourlas (2009) [19] | Cost of Quality Models and Their Implementation in Manufacturing Firms | This paper explains cost reduction is needed to achieve quality, and the reduction of these costs is only possible if they are identified and measured. | It was concluded that in order to improve quality, an organization must take into account the costs associated with achieving quality not only to meet customer requirements, but also to do it at the lowest possible cost. |
| 12 | R. Sampath Kumar and N. Alagumurthi (2009) [20] | Calculation of Total Cost, Tolerance Based on Taguchi's, Asymmetric Quality Loss Function Approach | The paper focuses on the manufacturing firms and the problems associated due to increase in quality costs. Quality loss function is studied in great detail. | It was concluded that for reducing the costs needed to achieve quality, a systematic focus on the problems should be given. The reduction of these costs is only possible if they are identified and measured. |

# Continued...

| Sr no | Author | Title | Tools & technique | Remarks |
| --- | --- | --- | --- | --- |
| 13 | Resit Unal, Edwin B. Dean (1991) [22] | Taguchi Approach to Design Optimization for Quality and Cost: An Overview | The paper discussed that Taguchi method is a powerful tool which can offer simultaneous improvements in quality and cost. The method can aid in integrating cost and engineering functions through the concurrent engineering approach required to evaluate cost over the experimental design. | The Taguchi method emphasizes pushing quality back to the design stage, seeking to design a product/process which is insensitive or robust to causes of quality problems. |
| 14 | Shyam Mohan (2002) [25] | Robust Design | The underlying principles, techniques& methodology of robust design are discussed in detail in this report with a case study presented to appreciate the effectiveness of robust design. The importance of Parameter design & Tolerance design as the major elements in Quality engineering are described. | The Quadratic loss functions for different quality characteristics are narrated, highlighting the fraction defective fallacy. |
| 15 | T. W. Simpson (2003) [26] | Taguchi's Robust Design Method | The paper proposed that many companies realized that the old methods for ensuring quality were not competitive with the Japanese methods. It is also stated that no amount of inspection can improve a product quality. | It was concluded that quality must be designed into a product from the start. Taguchi's robust design approach in an effort to improve product quality and design robustness was the only solution. |

# Continued...

| Sr no | Author | Title | Tools & technique | Remarks |
|---|---|---|---|---|
| 16 | V.G. Surange and S.N.Teli (2013) [27] | Effective Utilization of Quality Cost Reducing Tools in Automobile Industry | This paper carried out study on cost of poor quality in Indian automobile industries to overcome various related problems by using several cost of quality tools such are critical to quality and critical to process, Pareto analysis, SIPOC diagram, fishbone analysis, control charts, scatter plot, why-why analysis. | The various objectives are achieved by application of cost of quality tools to quality improvement project in automotive industry. |
| 17 | Vytautas Snieska and Asta Daunoriene (2013) [28] | Hidden Costs in the Evaluation of Quality Failure Costs | The underlying principles, techniques& methodology of robust design are discussed in detail in this report with a case study presented to appreciate the effectiveness of robust design. The importance of Parameter design & Tolerance design as the major elements in Quality engineering are described. | The Quadratic loss functions for different quality characteristics are narrated, highlighting the fraction defective fallacy. |
| 18 | Zillur Rahman and Faisal Talib (2010) [30] | Study of Optimization of Process by Using Taguchi's Parameter design approach | The paper proposed that many companies realized that the old methods for ensuring quality were not competitive with the Japanese methods. It is also stated that no amount of inspection can improve a product quality. | It was concluded that quality must be designed into a product from the start. Taguchi's robust design approach in an effort to improve product quality and design robustness was the only solution. |

# Continued...

❑ Literature Gap:

It is relevant to carry out the gap analysis so that we may be able to explain that our study is new and fresh to whatever work that has been presented further. The problem was defined by studying the literature survey and finding literature gap as for carrying out research.

- There were many papers that were presented on cost of quality and robust design in manufacturing industries. Broad area of use of robust design methodology is to strength consideration or process optimization. But this methodology has seldom been used in dimensional variation in gears which defines the gap of my research work.

- It is observed very less practical work is presented on dimensional tolerances of gear which ultimately decreases the quality of the product and its life time durability. Hence, reduction in cost of quality in terms of dimensional tolerances is studied further and a case study is evaluated for the same.

- Dimensional variation leads to poor quality of products manufactured which ultimately effects the overall cost and profit of the company. My research aim is to produce the gear with minimum tolerances or dimensional variation so that reduction in cost of quality by application of robust design is achieved.

# 3.  Research Methodology

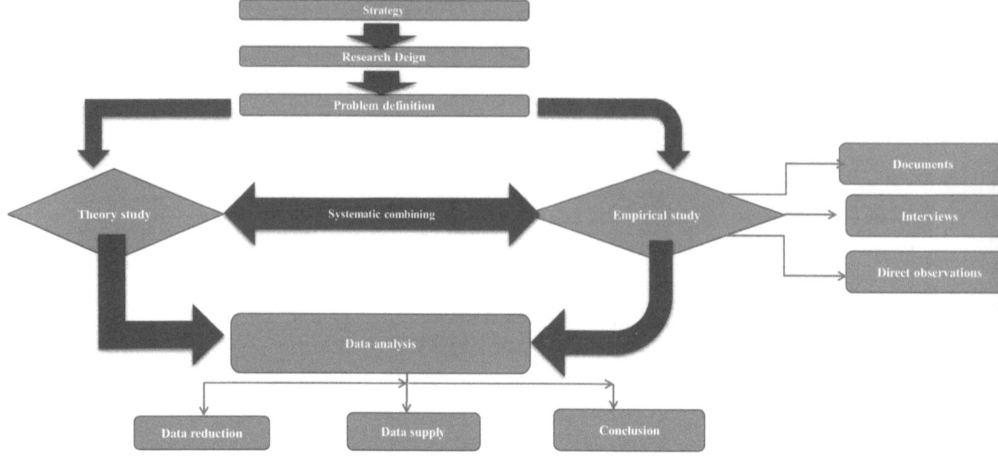


# Continued...

❏ Proposed Methodology:

- A research methodology is developed by implementing robust design methodology in particular taguchi quality loss function.

- COQ and Taguchi Design are integrated together for time and resource savings and determination of important factors affecting operation, performance and cost.

- A proposed methodology for research is as shown in the flowchart.

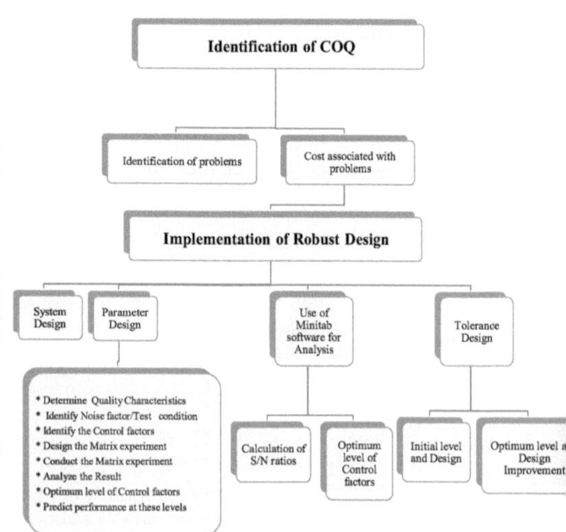

# Continued...

❑ Basic approach of Robust Design:

| | |
|---|---|
| Step 1 | • Identify main factors, side effects and failure mode |
| Step 2 | • Identify Noise factors, testing condition and quality characteristics |
| Step 3 | • Identify objective function to be optimized |
| Step 4 | • Select the orthogonal array matrix of experiment |
| Step 5 | • Conduct matrix experiment |
| Step 6 | • Analyze data and predict optimum level |
| Step 7 | • Perform verification of experiment |

# Continued...

❑ System Design at BGL:

- BGL produces various products that are manufactured to print.

- A hobbing machines and tool used for cutting gear teeth is by means of a hob.

- The hob is a cutting tool used to cut the teeth into the workpiece.

# Continued...

❑ Components and its Dimensions:

| Components | Material / Type |
|---|---|
| Spur gear | Mild Steel |
| Cutting tool | Hob cutter (Carbide tool) |
| Coolant | Oil based / Water based |

| Dimensions | Instruments |
|---|---|
| Root Diameter | Micrometer (0-25) 0.01 L/C |
| PCD (Pitch Circle Diameter) | Dial gauge (0-10) 0.01 L/C |
| Spline width | Micrometer (25-50) 0.01 L/C |
| Spline gauge | Spline ring gauge (GO gauge) |

# Continued...

❑ Quality losses at BGL:

| Prevention Costs | Appraisal Costs | Internal Failure Cost | External Failure Cost | Internal Supplier Cost | External Supplier Cost |
|---|---|---|---|---|---|
| -Checking Quality<br>-Finding good quality<br>-Time required for checking man M/C | -Inspection of product after production<br>-Feedback from customer<br>-Testing the quality | -Poor quality of machine<br>-cutting machine not working properly<br>-Use of low cutting tools or blades | -Unsatisfied customer demand or requirement<br>-Customer and suppliers will not invest in product | -Incorrect design<br>-Incorrect Material<br>-Damaged material | -Use of inefficient resources<br>-Quality loss by external supplier |

**COQ**

- Scrap/Rework
- Waiting time
- Replanning
- Delay/Overtime
- Start-up time
- Incomplete deliveries

# Continued...

❑ Parameter Design:

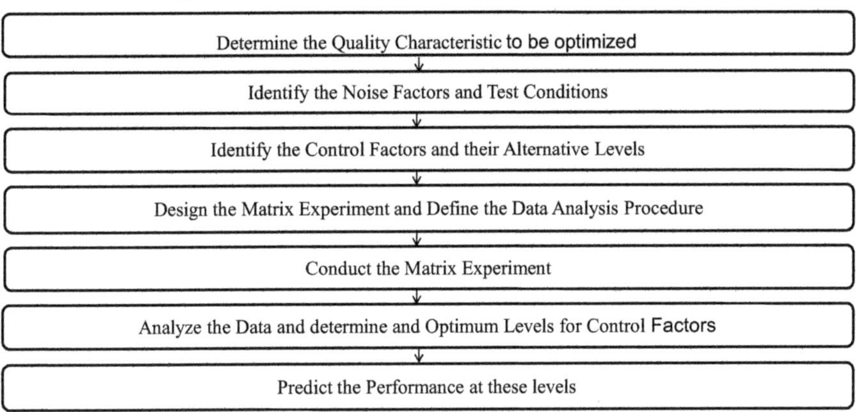


# Continued...

❑ Analysis by Minitab Software:

- Calculation of mean, Standard Deviation and Signal to Noise ratio
- Graphical representation for optimum value at various levels obtained from robust design approach

❑ Reduction in COQ:

- The cost of quality at previous level were studied
- The cost of quality at optimum levels obtained from robust design approach is also studied
- Comparison from the results are drawn

# 4.  Reduction in COQ by using Robust Design

❑ Case Study in Automobile Industry:

- The objective of this case study is to reduce Cost of Quality by using Robust Design in Bharat Gears Limited.
- Even though gear where within the quality range as per specification, it had many errors such as P.C.D error, lead error, etc. which affected the quality of the gears produced.
- Hence, the idea of implementing robust design to reduce the dimensional tolerance and thereby decreasing the cost of quality of a product.
- Various control factors such as hob speed, in feed and axial feed of tool were identified as control factors and levels of control factors were set.
- Discussion with the workers and engineers helped us determine the various noise factors such as vibration in hobbing machine, coolant used, grain structure of the gear material, initial casting methods used to produce blanks, etc.

# Continued...

❑ Problem Identification:

- The gears are produced by using the parameter value as shown below

| Parameters | Values |
|---|---|
| Hob Speed | 250 Rpm |
| In Feed | 0.5 mm |
| Axial Feed | 2.00 mm/rev |

The output response should be within given range so that the gear is regarded as of good quality

| Dimensions | Output Response (Tolerance) |
|---|---|
| Outer Diameter | 34.70- 34.77 mm |
| Root diameter | 27.90 - 28.10 mm |
| Spline Width | 8.59 - 8.62 mm |
| Spline Width Taper | 0.02 mm max. |
| P.C.D. Run out | 0.04 mm |
| Lead Error | 0.015 mm |
| Lead Error | 0.05 mm max. |

# Continued...

- We observed that the initial root diameter of spur gear was within the tolerance 27.90-28.10 mm has the highest effect on Cost of Quality in terms of tolerance design. Due to this high tolerance limit of root diameter of gear up to 0.2 mm many gears were having life less than one year.
- High tolerance leads to high rejection rate at working level.

❑ Identification of Cost of Quality:

- Inspection or measurement is classified as Conformance cost. The price of conformance is the cost involved in making certain things that are done right the first time

# Continued...

❑ Tolerance Design for Root diameter:

Root Diameter is the diameter of the Root Circle that passes through the bottom of the tooth spaces.

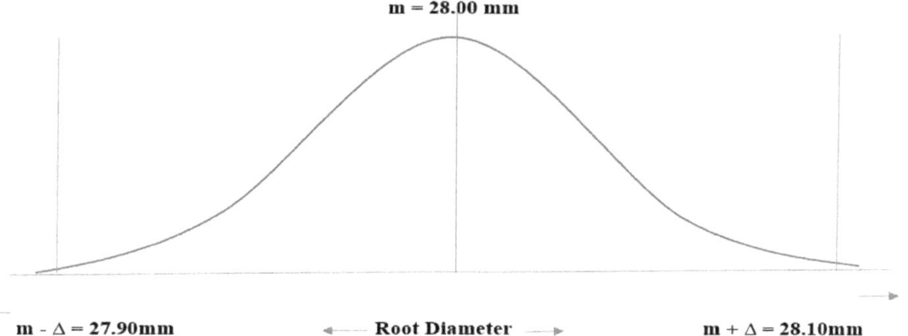

# Continued...

- All the gears with in the tolerance limit were initially accepted as rejections were not identified during 1st inspection.

- Whenever the product deviated from the target value of dimension 28.00 mm, the quality of the gear was indirectly detoriated.

- The traditional approach employed by manufacturers has been to ensure that performance fell within the upper and lower specification limits as shown in figure.

- Initially the gears failing with in the limit 27.90 mm – 28.10 mm were accepted which were sometimes of poor quality leading to quality loss and money loss.

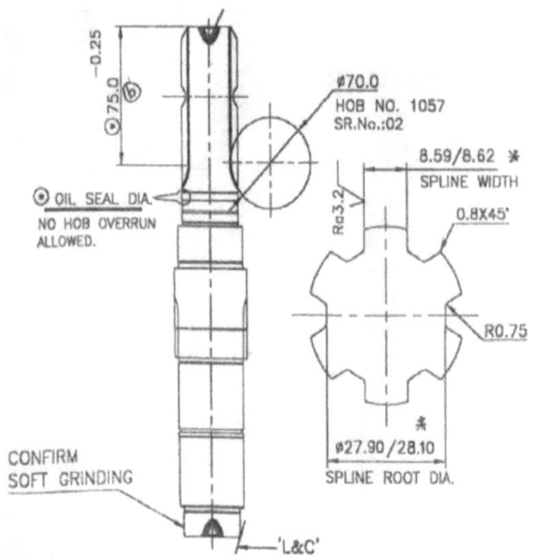


# Continued...

❑ Robust Design Methodology:

- **Quality Characteristics** - Dimension of Spur gear

Tolerance limit for Root Diameter of Spur gear is a quality characteristic to be determined.

- **Control factors** - Speed of Hobbing Machine in rpm

  In feed of cutting tool in mm

  Axial feed in mm/rev

| Control Factors | Level | | |
| --- | --- | --- | --- |
| | 1 | 2 | 3 |
| Speed (rpm) | 250 | 275 | 300 |
| In Feed (mm) | 0.4 | 0.5 | 0.6 |
| Axial Feed (mm/rev) | 1.9 | 2.0 | 2.1 |

# Continued...

- **Noise factors**      -      Vibration in hobbing machine
  -      Coolant used
  -      Grain structure of material

| Sr. No. | Noise factors | Causes |
|---|---|---|
| 1 | Vibration in Hobbing machine | Improper fitting of foundation bolt |
| | | Loose bearing in machine |
| | | Fixture foundation |
| | | Wobbling in cutting tool |
| | | Improper mounting of gear blank |

# Continued...

| Sr. No. | Noise factors | | Causes |
|---|---|---|---|
| 2 | Coolant used | Water based | These are used when gears are exposed for longer time or even when kept in storage due to rust formation. |
| | | | It is also used at shipping or package stage due to corrosion |
| | | | It is less heat absorbent |
| | | | It has better lubrication |
| | | Oil based | High cutting speeds can be obtained |
| | | | Rust protection even in packaging and transportation stages. |
| | | | Very less chances of corrosion |
| | | | Economical in overall cycle operation |

# Continued...

| Sr. No. | Noise factors | Causes |
|---|---|---|
| 3 | Grain structure of material | The grain structure of gear blank to be processed should be stream lined with the nature of cut. This help in cutting of gear splines properly. |
| | | Many a times such a structure is not presented which leads to various error in gear cutting profile |
| | | Also the cooling rate while cutting and heat dissipation depends upon the grain structure |

# Continued...

- **Test Condition**     -     3 Control factors & 3 Levels

| Experiment Runs | Speed (rpm) | In feed (mm) | Axial feed (mm/rev) |
|---|---|---|---|
| | Level 1 | Level 2 | Level 3 |
| 1 | 250 | 0.4 | 1.9 |
| 2 | 250 | 0.5 | 2.0 |
| 3 | 250 | 0.6 | 2.1 |
| 4 | 275 | 0.4 | 2.0 |
| 5 | 275 | 0.5 | 2.1 |
| 6 | 275 | 0.6 | 1.9 |
| 7 | 300 | 0.4 | 2.1 |
| 8 | 300 | 0.5 | 1.9 |
| 9 | 300 | 0.6 | 2.0 |

# Continued...

- **Conduct the Matrix Experiment**    -    Output response is Root Diameter (mm)

| Experiment Runs | Speed (rpm) | In feed (mm) | Axial feed (mm/rev) | Root Diameter (mm) | | | Output Mean (mm) |
| --- | --- | --- | --- | --- | --- | --- | --- |
| | Control Factors | | | | | | |
| | 1 | 2 | 3 | Y1 | Y2 | Y3 | |
| 1 | 250 | 0.4 | 1.9 | 27.93 | 27.95 | 27.96 | 27.94 |
| 2 | 250 | 0.5 | 2.0 | 27.92 | 27.94 | 27.97 | 27.94 |
| 3 | 250 | 0.6 | 2.1 | 27.92 | 27.96 | 27.99 | 27.95 |
| 4 | 275 | 0.4 | 2.0 | 28.05 | 28.07 | 28.09 | 28.07 |
| 5 | 275 | 0.5 | 2.1 | 27.99 | 27.99 | 28.00 | 27.99 |
| 6 | 275 | 0.6 | 1.9 | 27.94 | 27.93 | 27.90 | 27.92 |
| 7 | 300 | 0.4 | 2.1 | 27.98 | 28.01 | 28.04 | 28.01 |
| 8 | 300 | 0.5 | 1.9 | 27.91 | 27.94 | 27.92 | 27.92 |
| 9 | 300 | 0.6 | 2.0 | 28.04 | 28.06 | 28.09 | 28.06 |

# 5.    Results and Calculations

❑ Experimental Results:

- Minitab software is used to calculate the values for mean, standard deviation and S/N ratios and all the values are plotted graphically.
- The nominal value as in case of dimensional tolerances from S/N graph is selected as the optimum value.
- Minitab performs all the steps required for taguchi design for a designed orthogonal array.
- S/N ratio and means are calculated and various graph for analysis is drawn by using minitab 17 software.
- The S/N ratio for root diameter of spur gear is calculated on Minitab 17 Software using Taguchi Method.

# Continued...

- Values of mean, StdDev and S/N ratio is found out and plotted graphically as shown below

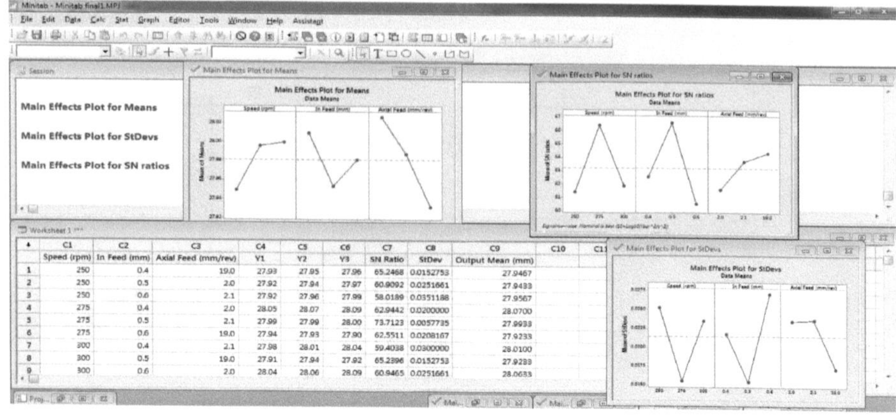

# Continued...

- Plot of mean and standard deviation graphically is shown below

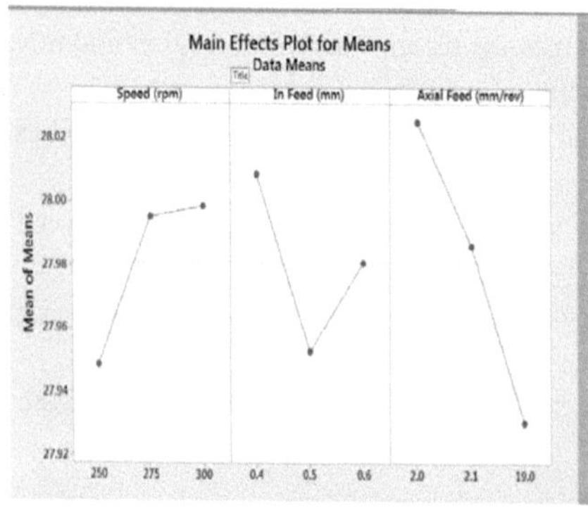

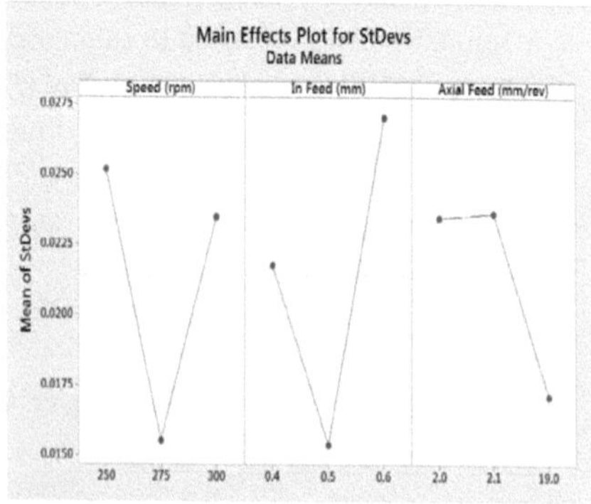

# Continued...

❑ Calculations:

- The S/N ratio is calculated by using the formula $S/N = 10 \times \log\left((Y^2) \div \sigma^2\right)$ , where Y is output mean and sigma is standard deviation. The various values for S/N ratio is as shown below

| Experiment Runs | Speed (rpm) | In feed (mm) | Axial feed (mm/rev) | Output Mean | S/N ratio |
| | Control Factors | | | (mm) | (Nominal best) |
| | 1 | 2 | 3 | | |
|---|---|---|---|---|---|
| 1 | 250 | 0.4 | 1.9 | 27.94 | 65.2468 |
| 2 | 250 | 0.5 | 2.0 | 27.94 | 60.9092 |
| 3 | 250 | 0.6 | 2.1 | 27.95 | 58.0189 |
| 4 | 275 | 0.4 | 2.0 | 28.07 | 62.9442 |
| 5 | 275 | 0.5 | 2.1 | 27.99 | 73.7123 |
| 6 | 275 | 0.6 | 1.9 | 27.92 | 62.5511 |
| 7 | 300 | 0.4 | 2.1 | 28.01 | 59.4038 |
| 8 | 300 | 0.5 | 1.9 | 27.92 | 65.2396 |
| 9 | 300 | 0.6 | 2.0 | 28.06 | 60.9465 |

# Continued...

- S/N ratio for Speed (rpm):

S/N ratio for level 1 is $SN_{L1}$ = (65.2468 + 60.9092 + 58.0189)/3 = 61.39

S/N ratio for level 2 is $SN_{L2}$ = (62.9442 + 73.7123 + 62.551)/3 = 67.07

S/N ratio for level 3 is $SN_{L3}$ = (59.4038 + 65.2396 +60.9465)/3 = 61.8633

| Speed (rpm) of Hob | Level | S/N Ratio |
|---|---|---|
| 250 | 1 | 61.39 |
| 275 | 2 | 67.07 |
| 300 | 3 | 61.86 |

It can be observed that the nominal value is for Level 3 i.e. Hob Speed of 300 rpm is one of the optimum parameter.

# Continued...

- S/N ratio for In feed (mm):

| In feed (mm) | Level | S/N Ratio |
|---|---|---|
| 0.4 | 1 | 62.53 |
| 0.5 | 2 | 66.62 |
| 0.6 | 3 | 60.50 |

It can be observed that the nominal value is for Level 1 i.e. In feed of 0.4 mm is one of the optimum parameter.

- S/N ratio for axial feed (mm):

| Axial feed (mm/rev) | Level | S/N Ratio |
|---|---|---|
| 1.9 | 1 | 61.59 |
| 2.0 | 2 | 63.71 |
| 2.1 | 3 | 64.34 |

It can be observed that the nominal value is for Level 2 i.e. Axial feed of 2.1 mm/rev is one of the optimum parameter.

# Continued...

❑ Optimum parameter:

- S/N ratio is selected as the parameter to study the dimensional effect in gear. Optimum value is to be selected in case of dimensional tolerances to give optimum or maximum output. To achieve optimum tolerance design for root diameter for spur gear we conclude the following results:

  o Hob speed at 275 rpm

  o In feed of 0.4 mm

  o Axial feed of 2.0 mm/rev

- Optimum parameter to achieve best dimensional tolerance in gear used in automobile thereby reducing the cost of poor quality so as to minimize the deviation from the target value.

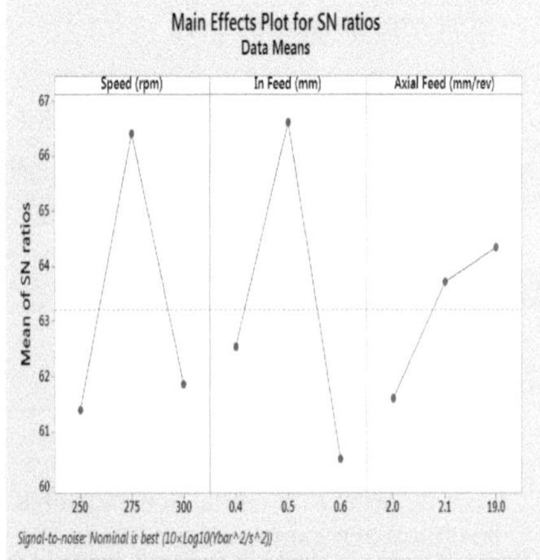

# Continued...

❏ Predict the performance at optimum level:

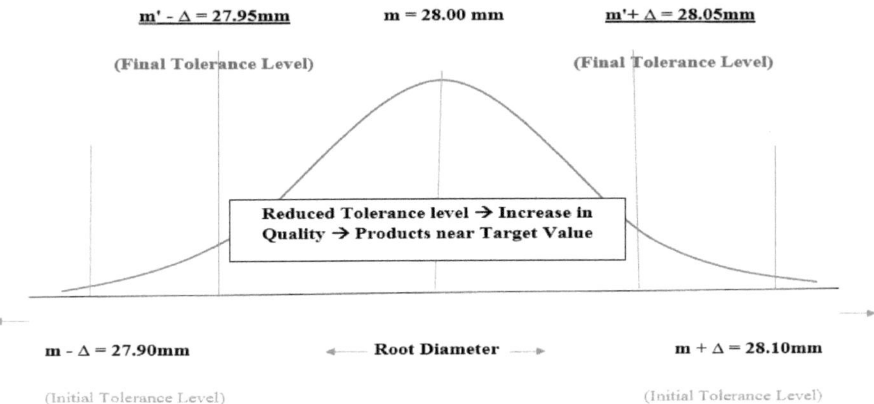


# Continued...

In the figure, m - $\Delta$ is lower limit of tolerance for gear and m $_+$ $\Delta$ is upper limit of tolerance for the gear before application of taguchi design whereas m' - $\Delta$ is lower limit of tolerance for gear and m' + $\Delta$ is upper limit of tolerance for gear. Following discussions were made from the results obtained:

- Experiments were carried out at optimum setting for a set of gears and their root diameter tolerance came within 27.95 – 28.05 mm.

- It meant that the tolerance was reduced from 0.2 mm to 0.1 mm.

- It exactly fits the Taguchi design as Taguchi relates quality at target value in robust design approach.

- It meant that the design was made more robust as it is near to the target value. Any deviation from the target is termed as quality loss which is reduced in this case.

- Our primary goal is to reduce the tolerance deviation so that the life of gear is increased thereby the wear and tear due to meshing with other will also get minimized.

# Continued...

❑ Reduction in COQ:

- A case study on manufacturing of gears which is used in automobile industry for transmission of power is studied by taking cost of quality in consideration.
- The spur gear are used along with transmission shaft in a gear box assembly for transmission of power.
- Various factors that affects the overall COQ are assembly cost, installation cost, manufacturing cost, warranty cost, etc. Out of these manufacturing cost and warranty cost are the main factors that are studied under the thesis.
- After finding optimum parameters for manufacturing of gears, several improvement in the process and overall reduction in cost of quality was achieved. A data for one month for COQ reduction is studied.

There are mainly two types of rejections namely,

- **Internal rejection** – Tolerance range for root diameter, Lead error, P.C.D error, etc.
- **External rejection** – Failure of gear at working stage, rejection due to improper mating of gear with transmission shaft, etc.

# Continued...

- Causes for Internal and External rejections:

| Internal Rejections | External Rejections |
| --- | --- |
| Scrap | Returned products |
| Rework | Customer complaints |
| Repairs | Field repair call |
| Unscheduled and unplanned services | Warranty expenses |
| Defect removal | Poor safety |
| Lost process time | Replacement |

# Continued...

☐ Traditional approach and Robust approach:

### Case I: COQ due to Traditional approach

1. In this case, internal rejections were low due to high tolerance limit of 0.2 mm.

2. When these gears were used in assembly with transmission shaft or during working stage many of the gears failed or produced hauls in the system. Hence, external rejections were high.

3. The overall brand of the company was affected and lead to loss of trust between customer and supplier.

4. Quality of the product is compromised leading to increasing in COQ of the product.

### Case II: COQ due to Robust approach

1. In this case, the tolerance limit was reduced from 0.2 mm to 0.1 mm. Hence, Internal rejections were high.

2. External rejections were low as majority of the gears were rejected internally.

3. These improved the brand image of the product and gear design were more robust as compared to traditional approach.

4. Quality cannot be compromised at any stage. Increase in quality lead to reduction in COQ of the product.

# Continued...

☐ Comparison of COQ:

A data of one month for COQ reduction is studied and explained below. We found the following estimation and changes in total quality cost.

| COQ | Reason for failure | Initial COQ (Rs.) | Final COQ (Rs.) |
|---|---|---|---|
| | | Case I | Case II |
| Internal Failure cost (Manufacturing cost, Machining cost) | Gear not within tolerance specifications, Improper cutting of gear, improper hob used for gear cutting, etc. | 25850 | 26550 |
| External Failure cost ( Loss at the hand of customer) | Working condition failure, improper installation with transmission shaft. | 37135 | 24400 |
| Total | | 62985 | 50950 |
| COQ Reduction | | 19.10 % | |

From this study, it is found that in the manufacturing department in which we made improvements total cost of quality is reduce by 19.10 %.

# 6.　Conclusion

- Robust design is implemented in a gear hobbing process. The optimum setting of parameter is to be obtained in presence of uncontrollable noise factors. The target value for root diameter of a spur gear used with transmission shaft in an automobile is 28.00 mm.

- Taguchi parameter design is used to find an optimum value so that component meets the desired specification or in other words meets the target value. Experimental analysis is performed at different levels of control factors and noise factors. Optimum value of control factors is found by studying Signal to Noise ratio (nominal the best).

- Gears are produced and it was effectively found that the tolerance range was found to be 27.95-28.05 mm. In other words, the tolerance was reduced from 0.2 mm to 0.1 mm which led to overall reduction in cost of quality of a spur gear.

- Robust design integrated with cost of quality approach leads to reduction in cost of poor quality. It is a powerful quality improvement and design improvement tool.

# Continued...

- It can be used for existing process improvement. It can be used for improvement in existing product design or in the process of a new development.

- Robust design is now largely used in Automobile manufacturing sector to make the process insensitive to variation. It helps in defect reduction, process improvement, tolerance improvement and reduction in cost of poor quality.

- Robust design and cost of quality are closely associated and supportive to each other. Tolerance design of a product is closely related to robust design and their mapping together leads to reduction of defective parts at operating stage.

- Robust design is a methodology that has been successfully applied in design stage and manufacturing stage. Products life cycle is increased by studying various parameters and finding their optimum value eventually affecting the product's performance.

# 7.  Papers Published

- "Reduction of Cost of Quality by using Robust Design: A Research Methodology", *International Journal of Mechanical and Industrial Technology (IJMIT)*, ISSN 2348-7593 (Online) Vol. 2, pp: (122-128), Month: October 2014 - March 2015.

- "Reduction of Cost of Quality by using Robust Design: A Case Study in Automobile Industry", *International Journal of Mechanical and Industrial Technology (IJMIT)*, ISSN 2348-7593 (Online) Vol. 2, Issue 2, pp: (109-116), Month: October 2014 - March 2015.

# 8.  References


- Achamyeleh A. Kassie, Samuel B. Assfaw, "Minimization of Casting Defects", *IOSR Journal of Engineering (IOSRJEN)*, 2002.

- Adil and A. Moutawakil, "The Quality Cost Reduction in Hollow Glass Manufacturing by Taguchi Method", *Journal of Scientific Research Publications*, Pg. 155-172, 2012.

- Andrea Schiffauerova, Vince Thomson, "A Review of Research on Cost of Quality Models and Best Practices", Department of Mechanical Engineering, *McGill University*, Montreal, Canada, 2007.

- B P Gautham, Nagesh Kulkarni, Janet K. Allen, "Robust Design of Gears with Material and Load Uncertainties", *IOSR Journal of Engineering (IOSRJEN)*, 2004.

- Davison Zimwara, Lameck Mugwagwa, Daniel Maringa, Albert Mnkandla, Linden Mugwagwa, Tendai Talent Ngwarati, "Cost of Quality as a Driver for Continuous Improvement-Case Study – Company X", *International Journal of Innovative Technology and Exploring Engineering (IJITEE)*, January 2013.

- Dasthagiraiah, N. V. Subba Reddy, N. Ramanamma, K. Harshavardhan Reddy, V. Naresh Reddy, K. Rama Devi, "Analyze Gear Failures and Identify Defects in Gear System for Vehicles Using Digital Image Processing", *IOSR Journal of Electronics and Communication Engineering (IOSR-JECE)*, 2013.

- Dobrin Cosmin, Stănciuc Ana-Maria, "Cost of Quality and Taguchi Loss Function", *Bucharest University of Economic Studies*, Romania.

- Fritz Klocke, Markus Brumm and Eva Gräser, "Overlap in Rolled PM Gears", *Metal Powder Industries Federation Publications*, 2009.

# Continued...


- G. Dennis Beecroft, Cost of Quality, "Quality Planning and the Bottom Line", *Bennis Beecroft Inc.*, 2000.

- Joseph Juran, "Quality Cost Analysis: Benefits and Risks", *Cem Kaner Publication*, January 1996.

- Joe Mangino, Jean-Pierre Fontelle "Quality Assurance and Quality control", Chapter 8, 2004.

- Juliana Litecka, Tomas Horvat & Veronika Fecova, "Classification of Factors affecting Hob wear", 2007.

- K.Gopinath & Prof. M.M.Mayuram, "Gear Manufacturing", *International Journal of latest Trends in Engineering and Technology (IJLTET)*, 2008.

- K R Balachandran and Bin Srinidhi, "Target Analysis: Cost, Quality or Both", *Stern School of Business*, New York University, April 1991.

- Krishankant, Jatin Taneja, Mohit Bector, Rajesh Kumar, "Application of Taguchi Method for Optimizing Turning Process by the effects of Machining Parameters", *International Journal of Engineering and Advanced Technology*, October 2012.

- Mahesh Krishna Shukla, Prakash Chandra Agrawal, "Impacts of Cost of Poor Quality in Indian Automobile Sector", *International Journal of Engineering Research and Applications (IJERA)*, April 2012.

- Mats Akerblom, "Gear Noise and Vibrations", *American J. of Engineering and Applied Sciences Publications*, 2012.

- Mohd. Muzammil, Prem Pal Singh, and Faisal Talib3, "Optimization of Gear Blank Casting Process by Using Taguchi's Robust Design Technique", *American J. of Engineering and Applied Sciences Publications*, 2003.

- N.M.Vaxevanidis, G. Petropoulos, J. Avakumovic, A. Mourlas, "Cost of Quality Models and Their Implementation in Manufacturing Firms", *International Journal for Quality research*, 2009.


# Continued...


- R. Sampath Kumar and N. Alagumurthi, "Calculation of Total Cost, Tolerance Based on Taguchi's, Asymmetric Quality Loss Function Approach", *American J. of Engineering and Applied Sciences Publications*, 2009.

- R. Ramesh, "Detection of local faults in rotating machines", *American J. of Engineering and Applied Sciences Publications*, 2012.

- Resit Unal, Edwin B. Dean, "Taguchi Approach to Design Optimization for Quality and Cost: An Overview", *Annual Conference of the International Society of Parametric Analysts*, 1991.

- Schiffauerova and Thomson, "A review of research on cost of quality models and best Practices", *International Journal of Quality and Reliability Management*, Vol.23, No.4, 2006.

- Shinn-Liang Chang, Jia-Hung Liu, Kai-Wei Jin, Ching-Hua Hung, and Shang-Hsin Chen, "Design Optimization for Robustness Considering the Gear Transmission Error", *Advance Online publication*, 2002.

- Shyam Mohan., "Robust Design," Department of Aerospace Engineering, *Indian Institute of Technology*, November 2002.

- T. W. Simpson, "Taguchi's Robust Design Method", *Concurrent Engineering*, 2003.

- V.G. Surange, Prof. S.N.Teli, Datta D. Adak, Siddhesh S. Rane, "Effective Utilization of Quality Cost Reducing Tools in Automobile Industry", *International Journal of Advanced Technology & Engineering Research (IJATER)*, March 2013.

- Vytautas Snieska, Asta Daunoriene, Alma Zekeviciene, "Hidden Costs in the Evaluation of Quality Failure Costs", *Inzinerine Ekonomika-Engineering Economics*, 2013.

- Y. Huang, S. Harte, L. Sud, John King, "Dynamic Parameter Optimization of an Automobile A/C Compressor Using Taguchi Method", *International Compressor Engineering Conference*, Purdue University Purdue e-Pubs, 1998.

- Zillur Rahman and Faisal Talib, "Study of Optimization of Process by Using Taguchi's Parameter Design Approach", Indian Institute of Technology, *The Icfai University Press*, 2010.

# YOUR KNOWLEDGE HAS VALUE

- We will publish your bachelor's and
  master's thesis, essays and papers

- Your own eBook and book -
  sold worldwide in all relevant shops

- Earn money with each sale

Upload your text at www.GRIN.com
and publish for free